FISH INVENTORY

CATEGORY	SPECIES	SWIMS ABOUT IN	RATING
ZAPPERS (PROBABLY BEST NOT TO TOUCH THEM)	ELECTRIC TORPEDOES	WORLDWIDE WARM WATERS	**FREAKY 5/5**
	EELS	FRESH WATERS LIKE THE AMAZON RIVER (THAT'S IN SOUTH AMERICA	
	CATFISH	FRESH WATERS IN AFRICA	
	STARGAZERS	THE ATLANTIC OCEAN	
STINGERS DEFINITELY DON'T TOUCH THEM	WEEVERFISH	THE ATLANTIC OCEAN	**STILL FREAKY 5/5**
	STINGRAYS	WARM BITS OF THE OCEAN	
	LIONFISH	THE SOUTH PACIFIC	
	SCORPION FISH STONEFISH }	THE INDIAN AND SOUTH PACIFIC OCEANS	
NOT MEMBERS OF A CIRCUS (AS FAR AS WE KNOW) →	CLOWNFISH	THE PACIFIC AND INDIAN OCEANS	
SINGERS (KINDA LIKE UNDERWATER KARAOKE)	OYSTER TOADFISH	MUDDY WATERS EG. SWAMPS	**FUNKY DUDES 4/5**
	CROAKING GOURAMI	STILLWATER LIKE PONDS OR CANALS OR PADDY FIELDS	
	BATFISH	THE PACIFIC OCEAN	

FREAKY, FUNKY FISH

Odd Facts about Fascinating Fish

written by
DEBRA KEMPF SHUMAKER

RP | KIDS
PHILADELPHIA

Illustrated by
CLAIRE POWELL

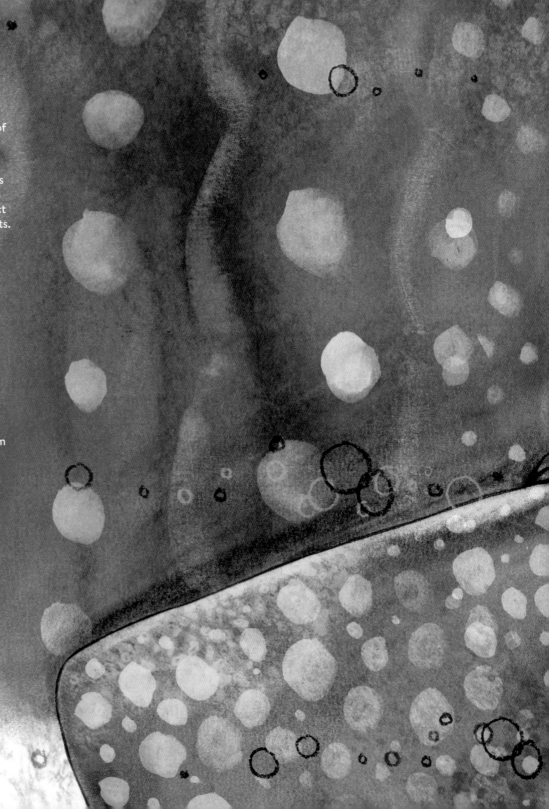

Running Press Kids
Hachette Book Group
1290 Avenue of the Americas, New York, NY 10104
www.runningpress.com/rpkids
@RP_Kids

Printed in China

First Edition: May 2021

Published by Running Press Kids, an imprint of Perseus Books, LLC,
a subsidiary of Hachette Book Group, Inc. The Running Press Kids name
and logo is a trademark of the Hachette Book Group.

The Hachette Speakers Bureau provides a wide range of authors for
speaking events. To find out more, go to www.hachettespeakersbureau.com
or call (866) 376-6591.

The publisher is not responsible for websites (or their content) that are not
owned by the publisher.

Print book cover and interior design by Frances J. Soo Ping Chow

Library of Congress Control Number: 2019955704

ISBNs: 978-0-7624-6884-3 (hardcover), 978-0-7624-6882-9 (ebook) ,
978-0-7624-7318-2 (ebook), 978-0-7624-7319-9 (ebook)

1010

10 9 8 7 6 5 4 3 2 1

To Jack, Sam, and Ben, for being my daily inspiration,
and to Tom, for everything.
—DKS

To Sophia and Bob, thank you for rescuing me
when I was lost at sea all those years ago.
—CP

Fish have **fins** and **gills** and **tails.**

WHALE SHARK

SAWFISH

PARROT FISH

CATFISH

COWHORN

SAILFISH

TORPEDO RAY

BLUESPOTTED RIBBONTAIL RAY

UNICORN FISH

RATTAIL

WEEVER FISH

MANTA RAY

SAND TIGER SHARK

COWTAIL STINGRAY

LIONFISH

HAMMERHEAD SHARK

STICKLEBACKS

GIANT OARFISH

SHEEPSHEAD FISH

OCEANIC WHITETIP SHARK

BIG EYED CARDINALFISH

ELECTRIC EEL

FLYING FISH

SUNFISH

SEAGRASS FILEFISH

ANGLERFISH

All fish **swim**
and most have **scales**.

STARGAZER

BATFISH

HAGFISH

PORCUPINEFISH

ZEBRA BATFISH

But...

ELECTRIC TORPEDO

ELECTRIC CATFISH

some fish **zap**

ELECTRIC EEL

ELECTRIC STARGAZER

SCORPIONFISH
OOOOOUCHH

STINGRAY
OWWWWWWW

and some fish sting.

STONEFISH
OOOHHHHHH

LIONFISH
EEEEEKK!

WEEVERFISH
AARGHHHH

FREAKY RATING | 5/5

CLOWNFISH

Did you know
that some fish sing?

OYSTER TOADFISH

CROAKING GOURAMI

FUNKY RATING: 4/5

BATFISH

One kind gives a **shake** and **swat**.

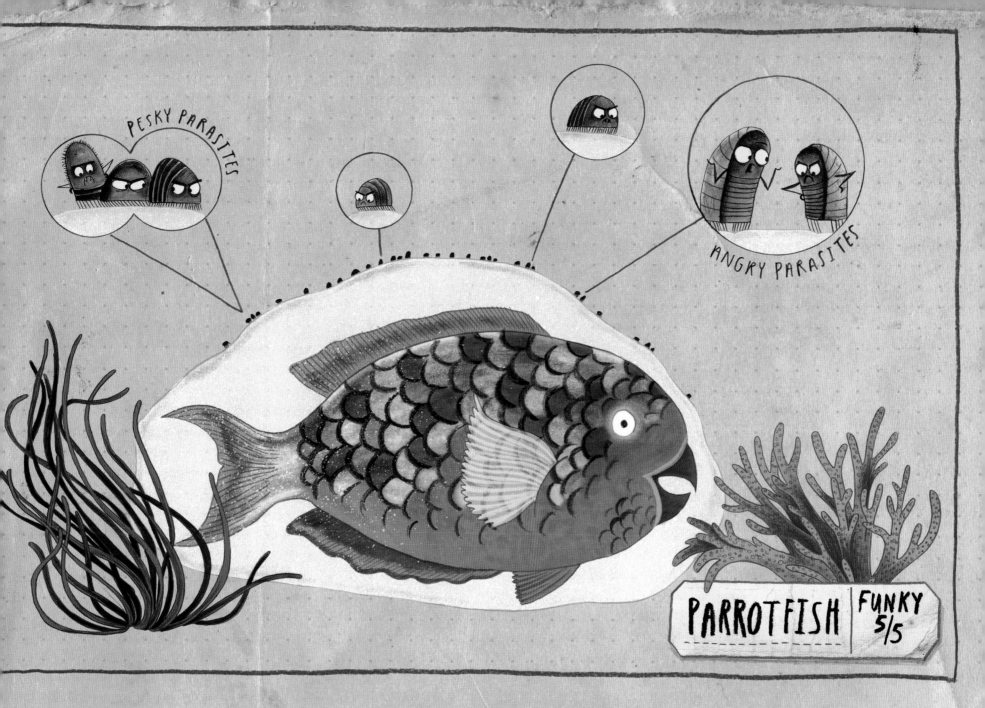

One fish coats itself with snot.

Some fish **dance**

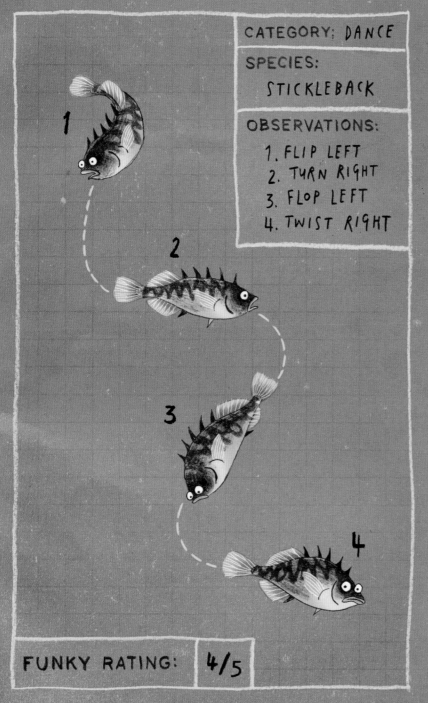

CATEGORY: DANCE

SPECIES:
STICKLEBACK

OBSERVATIONS:
1. FLIP LEFT
2. TURN RIGHT
3. FLOP LEFT
4. TWIST RIGHT

1

2

3

4

FUNKY RATING: 4/5

PREDATOR

| CICHLID | FUNKY RATING | 4/5 |

and some play **dead.**

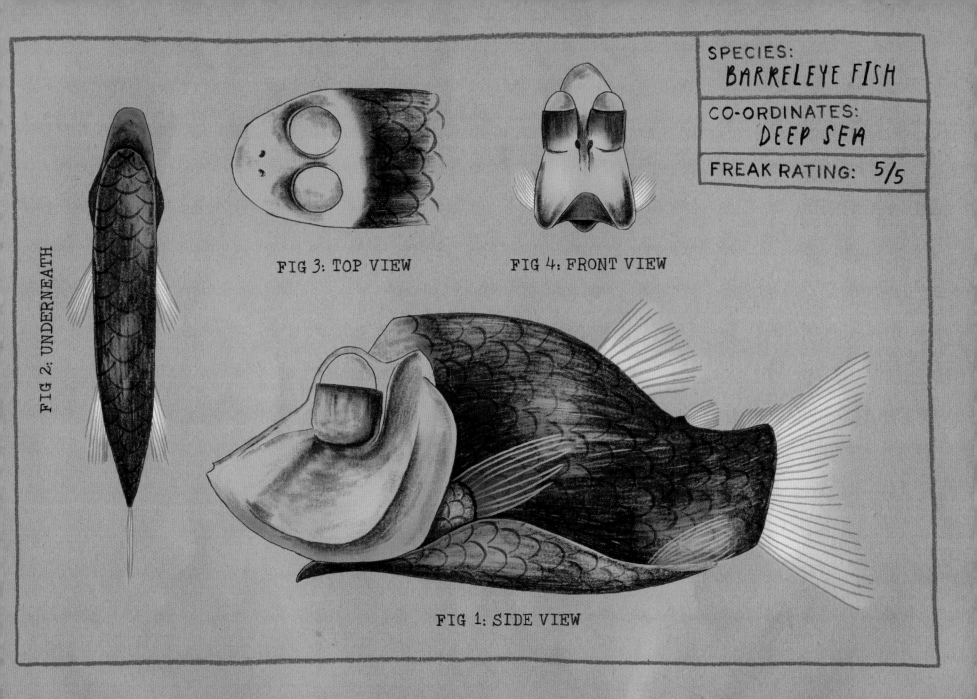

SPECIES:
BARRELEYE FISH

CO-ORDINATES:
DEEP SEA

FREAK RATING: **5/5**

FIG 2: UNDERNEATH

FIG 3: TOP VIEW

FIG 4: FRONT VIEW

FIG 1: SIDE VIEW

One fish sports a **see-through head!**

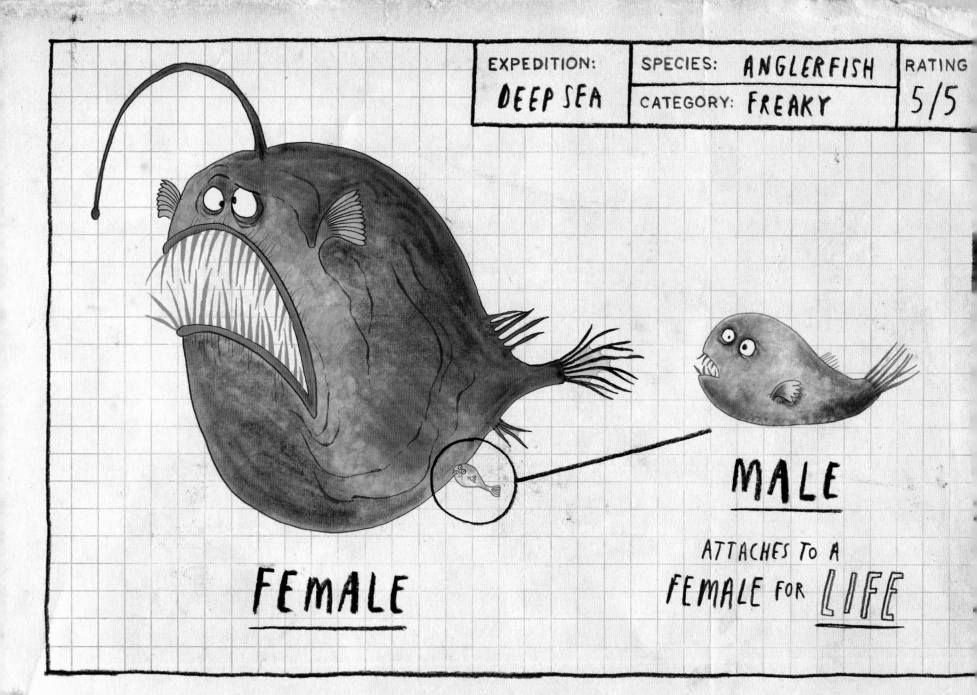

One gets **old** and starts to **shrink**.

Some have **lights** that they can **blink**.

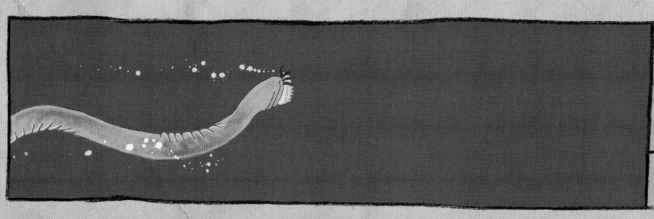

SPECIES: HAGFISH

FIELD NOTES:
PRODUCES SLIME
VERY QUICKLY

RATING:
FREAKINESS 5/5

[FIG 1]

[FIG 2]

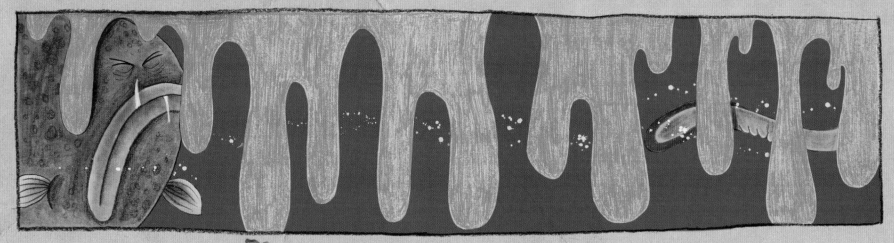

[FIG 3]

One fish squeezes out some slime.

Did you know that some fish climb?

Some fish **squirt**

ARCHERFISH | FUNKINESS ³/₅

PORCUPINEFISH

[FIG 1]

[FIG 2]

[FIG 3]

FUNKY RATING ⁵/₅

and some **inflate.**

AFRICAN LUNGFISH

FUNKY RATING 4/5

[FIG 1]

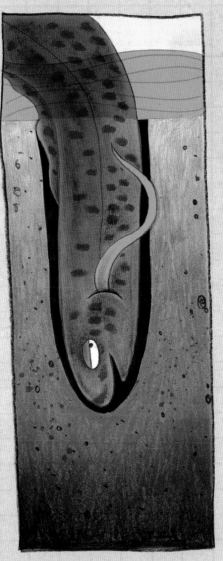

[FIG 2]

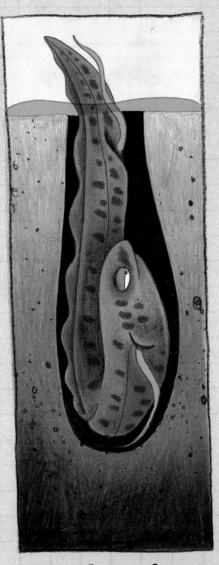

[FIG 3]

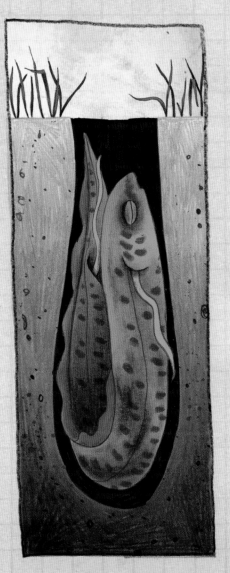

[FIG 4]

One fish needs to hibernate.

Some fish fly

FLYING FISH | PREFER TROPICAL AND SUBTROPICAL WATERS | FUNKY RATING 4/5

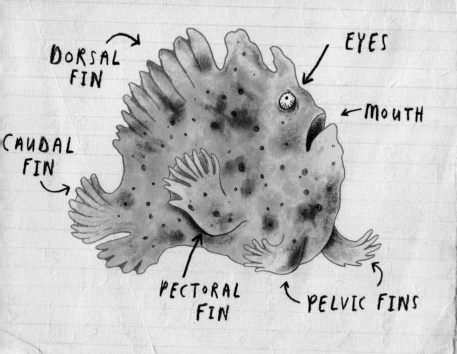

DORSAL FIN

EYES

MOUTH

CAUDAL FIN

PECTORAL FIN

PELVIC FINS

CATEGORY: **CREEPERS**

SPECIES: **CORAL REEF FROGFISH**

OBSERVATIONS: **HARD TO FIND**

FREAK RATING: **5/5**

SEA ROBIN
FUNKINESS
3/5

and some fish **creep.**

TIGER SHARK

BLOTCHED FANTAIL STINGRAY

OCEANIC WHITETIP SHARK

WHALE SHARK

SALMON SHARK

EAGLE STINGRAY

Some fish swim while they're **asleep.**

GREAT WHITE SHARK

BLUE SPOTTED RAY

DOGFISH SHARK

BAT RAY

LEOPARD RAY

YELLOW STINGRAY

HAMMERHEAD SHARK

THRESHER SHARK

ZEBRA SHARK

BLACKTIP SHARK

MANTA RAY

FUNKY RATING: 4/5

Funky ways to **stay alive.**

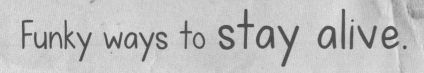

PORCUPINEFISH

STUPID ~~UNLUCKY~~ PREDATOR

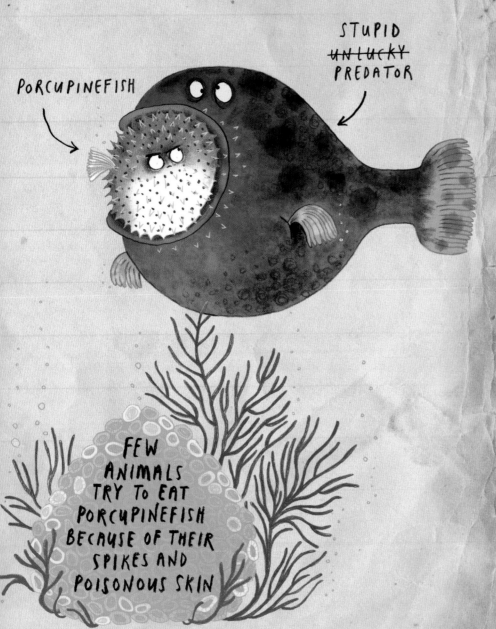

FEW ANIMALS TRY TO EAT PORCUPINEFISH BECAUSE OF THEIR SPIKES AND POISONOUS SKIN

[BROWN WATER SNAKE] →

CATFISH

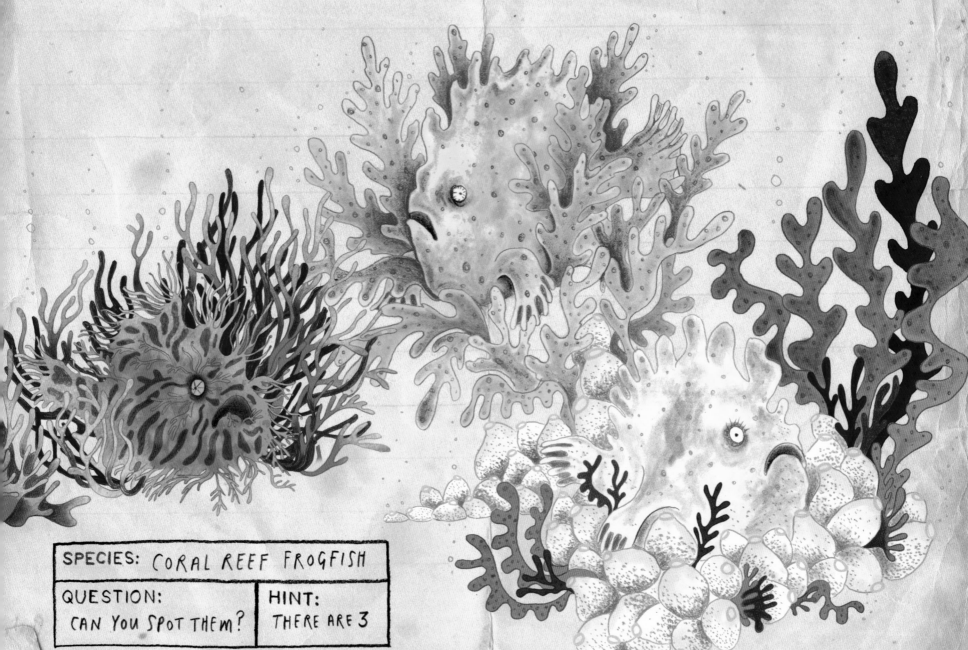

SPECIES: CORAL REEF FROGFISH

QUESTION: CAN YOU SPOT THEM?

HINT: THERE ARE 3

So many fish,
so many ways
a freaky fish
can spend its days.

EXPEDITIONS ZAP + STING

EXPEDITION SING

WHO ARE THESE FREAKY FISH?

There are more than 32,000 species of fish! While all live in water, have fins and a tail, and use gills to breathe, many, many fish do not act or look the way we think fish "typically" do. Instead, they have evolved odd shapes or interesting ways to move or eat to survive in the varied environments they live in.

Here are just some of the many types of amazing fish that can be found in water all over the world:

FISH THAT ZAP: About 250 species of fish can make electricity with their bodies. They use the electricity to both defend themselves and catch prey. The electric torpedo, catfish, eel, and stargazer produce a powerful zap that can kill small fish and stun large ones. And yes, even you would feel the shock if you touched the South American electric eel. Most zapping fish live in the waters of Africa and South America.

FISH THAT STING: Some species of fish sting, including weeverfish, stingrays, lionfish, scorpionfish, and stonefish. They typically use their venom to stun prey or to defend themselves.

FISH THAT SING: While many people think fish are quiet—after all, does your pet goldfish make any noise in its bowl?—scientists have learned that many fish grunt, pop, croak, bark, hum, whistle, chirp, hoot, and buzz—all without vocal cords. Instead, they use different parts of their bodies to make noise—squeezing muscles, rubbing bones, even changing the way they swim—similar to what you do when you cluck your tongue or snap your fingers. Fish make noise to find mates, defend territories, keep schools together, etc. Frequently, these sounds combine together like a beautiful chorus, especially at dawn and dusk. Numerous species make noise, including clownfish, oyster toadfish, croaking gourami, and batfish.

FISH THAT SHAKE AND SWAT:
Hammerhead sharks live in warm areas of the ocean in large schools of more than 500 sharks. When a female is ready to mate, she shakes and swats her head from side to side to get the attention of the male sharks.

FISH THAT COATS ITSELF WITH SNOT:
At night, parrotfish will wrap themselves in snotty cocoons they cough up from their gills as they settle in to rest in coral reef crevices. Scientists think they do this to hide their smell from predators or to protect themselves from tiny, nighttime bloodsucking pests.

FISH THAT DANCE:
Sticklebacks are tiny fish that live in both coastal saltwater and fresh ponds, lakes, and rivers. When a male stickleback is ready to mate, his pale belly changes to a bright orange-red color and he builds a nest. He then performs a zig-zag dance in front of the nest to attract a female to lay eggs in it. He even guards the nest and looks after the young when they hatch.

FISH THAT PLAY DEAD:
Some fish play dead to lure their prey. Both the yellow jacket and Livingston's cichlid will sink to a lake bottom and lie on their side motionless. Their spotty coloration helps them look like decaying fish. As other fish come to investigate and feast—chomp!—that fish becomes the cichlid's dinner. Other fish play dead to avoid being dinner. The seagrass filefish is shaped like a large leaf. When frightened, it will play "dead" by floating motionless on its side.

FISH WITH A SEE-THROUGH HEAD:
The barreleye fish lives in the very deep and dark waters of the Atlantic, Indian, and Pacific Oceans. It has a transparent dome-shaped head filled with a clear, protective liquid surrounding two tube-shaped eyes. These eyes typically point upward but will rotate down to look straight ahead as it follows and eats its prey.

FISH THAT SHRINK:
Some species of the male anglerfish are tiny—much smaller than the female anglerfish—and when a male finds a mate, it bites her and stays on. Their tissues fuse together and the male gets everything he needs to survive in the deep, dark ocean. As he ages, he gets smaller and smaller, dissolving into the female until he is just a sac of some tissue and organs on her belly.

FISH THAT BLINK:
About two-thirds of the fish that live near the dark ocean bottom have light-producing structures in their body called photophores. One such fish, the flashlight fish, has a light organ under each eye. It uses this light to find prey when it swims up to shallower waters at night looking for food. The light actually comes from bacteria living in sacs under the eyes of the fish, which it turns off by rolling the light organ under a flap of skin. At times, flashlight fish blink these lights every few seconds, looking like fireflies in the ocean.

FISH THAT SLIME:
Hagfish have more than 100 slime glands lining their tubular body. When they are attacked, they will squeeze out a tiny amount—less than a quarter teaspoon—of slime. In less than half a second, that tiny amount mixes with seawater and expands by 10,000 times (about 4 cups worth)! This slimy mixture clogs the gills of the attacker so the hagfish can make its escape. Though the slime looks gooey and sticky, it is actually very soft.

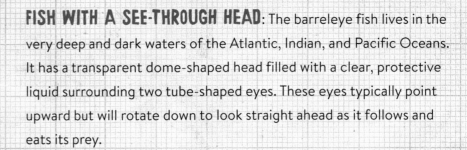

FISH THAT CLIMB: Mudskippers live in swampy forests of mangrove trees. They use their pectoral fins to climb roots and branches in search of food, spending much of their time out of the water. They store extra water in their gills so they can breathe for extended periods of time. When the oxygen in the stored water runs out, they go back into the water for more.

FISH THAT SQUIRT: The archerfish, which lives in freshwater rivers in India, Asia, and Australia, uses its squirting ability to hunt. When this fish spots a bug on vegetation above the water, it squirts a mouthful of water to knock down the bug and then gobble it up. Its squirt is so accurate and so strong that it can knock down a moving insect or one perched 4 feet away!

FISH THAT INFLATE: Porcupinefish have lots of spines on their body that lie flat as they swim around. But when they feel threatened, they will suck in water and puff out to a prickly ball to scare away the predator. It had better work, because they can't swim fast when they are inflated. Porcupinefish live in tropical oceans.

FISH THAT HIBERNATE: The African lungfish has both gills and lungs and lives in shallow pools of water. When the sun starts to dry up the pools, the lungfish burrows down into the mud and secretes a sticky mucus around itself like a cocoon. Once the water evaporates, the cocoon dries on the outside but keeps the fish moist and comfortable inside. It breathes air through a tube. Like mammals hibernating, the lungfish's metabolism slows down so it can survive several months or even years until it rains again. This long sleep during dry periods is called estivation.

FISH THAT FLY: Flying fish don't *actually* fly; they just look like they do. When flying fish are being chased by predator fish, they leap out of the water and use their large, wing-like pectoral fins to glide (not flap) through the air. As they descend, they move their tail back and forth to glide some more. Most can "fly" about 65 to 80 feet, but some glide as far as 650 feet before landing back in the ocean.

FISH THAT CREEP: Several species of fish creep on the bottom of the ocean floor. Sea robins, or gurnards, use their first three pectoral fins like feet to creep along the sea bottom. These "feet" have taste buds that sense small worms, crabs, and sand eels to eat. The coral reef frogfish has fins that are short and stout. They use them like legs to creep along the sea floor. They are so well camouflaged that they just crouch down to suck passing fish into their mouths.

FISH THAT SWIM WHILE ASLEEP: Most fish have swim bladders—a bag-like organ that holds gas from the fish's blood—that keep them afloat while resting. Sharks and rays, however, do not. Without a swim bladder, some species must swim at all times to avoid sinking, even while they are asleep. If some sharks stopped swimming, water wouldn't flow through their gill slits and they would not get enough oxygen to survive.

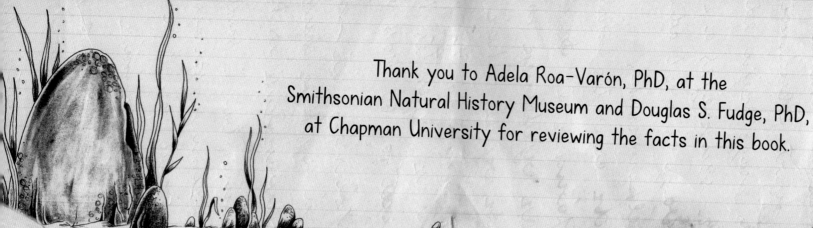

Thank you to Adela Roa-Varón, PhD, at the
Smithsonian Natural History Museum and Douglas S. Fudge, PhD,
at Chapman University for reviewing the facts in this book.

FURTHER LEARNING

To read more about these (and so many other) amazing fish, read the following:

Dipper, Frances. *Extraordinary Fish*. New York: DK Publishing, 2001.

Head, Honor. *Amazing Fish, Amazing Life Cycles*. New York: Gareth Stevens Pub, 2008.

Kalman, Bobbie, and Allison Lari. *What is a Fish? The Science of Living Things*. New York: Crabtree Publishing, 1999.

World Book Encyclopedia, Inc. *Animal Lives Series: Fish*. Chicago, IL: World Book, 2009.

SEE AND HEAR SOME OF THESE FREAKY, FUNKY FISH IN ACTION HERE:

"Armored Searobin: 2017 American Samoa," February 21, 2017, https://www.youtube.com/watch?v=B2Aeck7IKNs.

Crew, Bec. "Fish Have Been Recorded Singing a Dawn Chorus Just like Birds," September 23, 2016,
https://www.sciencealert.com/fish-have-been-recorded-singing-a-dawn-chorus-just-like-birds.

"Hagfish predatory behaviour and slime defence mechanism," October 26, 2011, https://www.youtube.com/watch?v=F8aVgSIDJjM

"Researchers Solve Mystery of Deep-Sea Fish with Tubular Eyes and Transparent Head," n.d.,
https://www.mbari.org/barreleye-fish-with-tubular-eyes-and-transparent-head/.

"Wonderful cichlids: Playing Dead," February 4, 2015,
https://www.pbs.org/video/earth-new-wild-wonderful-cichlids-playing-dead/.

SELECTED SOURCES

Caryl-Sue. "She's a Catch! Bizarre or beguiling? The weird, wonderful life of anglerfish," National Geographic online. https://www.nationalgeographic.org/media/shes-catch/

"Fish." *World Book Encyclopedia*, 2018.

Fudge, Douglas. Schmid College of Science and Technology, Chapman University. Email exchange with author, April 2019.

Helfman, Gene S. and Bruce B. Collette. *Fishes: The Animal Answer Guide.* Baltimore: Johns Hopkins University Press, 2011.

Knowlton, Nancy and Amanda Feuerstein. "No Fouling Around," December 2010. https://ocean.si.edu/ocean-life/fish/no-fouling-around.

Nelson, J.S., Grande, T.C., M.V.H. Wilson. *Fishes of the World*, fifth ed. Hoboken, NJ: John Wiley and Sons, 2016.

Owen, James. "The Living Dead: Animals That Pretend to Go Belly-Up," National Geographic, online. October 19, 2015. https://news.nationalgeographic.com/2015/10/151019-playing-dead-frog-possum-toad-animals-behavior-science/.

"Porcupinefish." *World Book Encyclopedia*, 2018.

Roa-Varón, Adela. National Museum of Natural History, Division of Fishes. Email exchange with author, May 2019.

Robison, Bruce and Kim Reisenbichler. "Macropinna microstoma and the Paradox of Its Tubular Eyes," *Copeia* 2008, No. 4, 780-784.

Yong, Ed. "No One Is Prepared for Hagfish Slime," *The Atlantic* online. January 23, 2019. https://www.theatlantic.com/science/archive/2019/01/hagfish-slime/581002/.

Yong, Ed. "Parrotfish sleep in a mosquito net made of mucus," National Geographic online. November 17, 2010. https://www.nationalgeographic.com/science/phenomena/2010/11/17/parrotfish-sleep-in-a-mosquito-net-made-of-mucus/.

FISH INVENTORY (CONT'D)

CATEGORY	SPECIES	SWIMS ABOUT IN	RATING
SHAKERS	HAMMERHEAD SHARKS	WARM TROPICAL BITS (SOUNDS NICE)	FREAK ULTIMO 5/5
SNOTTY (GIVE THEM A TISSUE)	PARROTFISH	TROPICAL AND SUBTROPICAL OCEAN PARTS	TOTALLY FUNKY 5/5
DANCERS	STICKLEBACKS	THE NORTHERN HEMISPHERE (EVERYTHING NORTH OF THE EQUATOR)	FUNKS GIVEN = 4
DEAD NOT REALLY THEY'RE JUST PRETENDING	CICHLIDS	CENTRAL AMERICA AND AFRICAN FRESHWATER	4 OUT OF 5
SEE-THROUGH	BARRELEYE FISH	THE ATLANTIC, PACIFIC AND INDIAN OCEANS	ER, MAXIMUM FREAK POINTS 5/5
OLD	ANGLERFISH	THE ATLANTIC AND THE ANTARCTIC DEEP DOWN	VERY FREAKY 5/5
BLINKERS	FLASHLIGHTFISH	THE PACIFIC AND INDIAN OCEANS	SUPER FUNKY 5/5
SLIMERS	HAGFISH	DEEP SEA	UNBELIEVABLY FREAKY 5/5
CLIMBERS	MUDSKIPPERS	COASTAL WATERS OF THE PACIFIC AND INDIAN OCEANS	FREAKS TO THE RATING 4/5